# DEAL & WALMER
# FROM THE GR...

## Martin Tapsell

## Tivoli Publishing

**67 York Road, Walmer, Deal CT14 7ED**

First Published by Tivoli Publishing 2011

Copyright   Martin Tapsell

All rights reserved. No part of this publication may be reproduced, stored in a retrieval system or transmitted in any form or by any means, electronic, mechanical, photocopy, recording or otherwise, without the prior consent of the publisher

ISBN  978-0-9560954-1-1

Printed and bound in Great Britain by Geerings Print, Ashford, Kent

Also by Martin Tapsell : Kent Cinemas revisited

Cover design by John Benzing

Cover Photographs : Gulley cleaning in Beach Street, Deal and on the back cover :  No stopping near James Robb Scott's modernist signal box (1938) in Western Road.

## INTRODUCTION

The Romans using their road from Dover to Richborough, are recalled by a history board in Great Mongeham, but the inland settlements clustered around St Leonards Church (1180) and Old St Marys Church, Walmer (1120) came much later. The name Deal derives from Addelam - latin Ad meaning at and Delam meaning a valley. There is no obvious valley now, but the High Street is still lower than the sea as flood pictures reveal, and the term Sea Valley is found in histories. Wealemere is Old English for the Britons' pond. The coastal area was colonised by fishermen in the early 16th century, initially with storehouses, windmills and chapels. Two Defensive earth bulwarks commenced c 1510 followed by the castles at Sandown, Deal, and Walmer in 1539-40, and all these required linking. Under Charles II, three roads were laid out parallel to the shore, Beach St., Middle St and Lower St (now High Street) with Lower Deal initially centred around the sea end of Queen St. The shanty town known as The Waste was placed under church management for nearly two centuries. War with France from 1792-1815 brought about more development of the Gosse (centred around Robert St) and the Outgrounds ( around Beaconsfield and Blenheim roads) Walmer benefited from a turnpike road and three barrack complexes, which were followed by the Regency villas built along the Beach. London Road was developed early in Queen Victoria's reign, one terrace being dated 1851. Fire, police and ambulance services are located on or just off London Road.

The closure of the naval yard, brought the completion of Victoria Town in the late 1860s . Mill Road developed from Edwardian times, but Mill Hill was built upon to house mining families from the early 1930s. Meanwhile, a few modernist houses were built at the far north of Deal, looking towards the golf courses.

A major change came in the Second World War when 173 high explosive bombs, 127 shells, 118 incendiaries and two parachute mines wrecked havoc. Out of 6,302 dwellings, 170 were destroyed, 1,599 badly damaged and 3,195 damaged to varying degrees. The Middle Street car park marks the most altered landscape. Much infill resulted over a wide area. A threat to Deal's historic quarter was finally averted by the designation of Kent's first conservation area in 1968. The withdrawal of the Royal Marines in 1996 has given an opportunity for many new homes and therefore roads, rather than new employment or extra supermarkets, whilst the Gas Board land around Cannon St becomes residential with a Community Centre opened, ahead of the curve, in July, 2011, next to Out Downs.

This boook will cover Deal and Walmer, not Great Mongeham or Sholden, hence the exclusion of for example Red House Wall Lane. The Deal/Walmer boundary is often the railway line, but at its north eastern end it bisects Deal Castle! Street names did not always have official origin. Thus in 1880 Walmer Board officially minuted the names and extent of the existing roads such as that Dover Road stretched from the Drum Inn to the parish boundary. The climate may be mainly dry, but gulley cleaning, as shown on the cover, is always in demand, whilst refuse disposal faces a wheelie bin revolution this autumn. The perception that Deal is a quiet backwater is weakened by attempts to cross the busy Strand or Dover Road, and by the first High Speed trains stopping here on September 5th. In August 2011 our MP., Charlie Elphicke is asking – Do you want a dual carriageway and more economic prosperity, or a peaceful retirement Balamorey ?

We are concerned with the streets of Deal and Walmer today, rather than how they looked like in the past, or with individual buildings. The town has much to offer that is not commonplace, including a pier, two Tudor castles and the Timeball tower, and has Kent's first conservation area, so has a broad range of buildings of interest, many of them listed. Public transport is adequate, provided subsidies are not withdrawn during the austerity years. The population is socially mixed. This is reflected in the diverse people who have given their names to the streets, or who were portrayed in a photographic exhibition at the Astor theatre.

P.S. Purists may deplore the loss of the apostrophe on road signs but I just record it.

Martin Tapsell, August, 2011

# GAZETTEER OF DEAL AND WALMER STREETS

CODES USED   A – In 1898 street directory;  B – on a Bus route, 2011;  C – Conservation Area;  CDC – Captain of Deal Castle; D– Descriptive or directional  H - Alludes to an historical event  or building;  LW – Lord Warden;   M – A former Military  connection or site;  PH – Recalls public house or Hotel,   W - A  Walmer street.

**ALBERT ROAD** (B)  Recalls Queen  Victoria's consort, 1819-61 The East Kent bus garage was located here from 1909-81 near Build Center.  Off Albert Road is **The Garth**, a Norse word meaning secluded garden

**ALFRED ROW**    From   Alfred  Square  (qv)  to  Enfield St. Previously known as The Drene, as an open  drainage channel ran along the row.

**ALFRED SQUARE** (A,C)  Probably after the Saxon  King Alfred but Alfred House  Academy,  later Deal College stood here  from 1800 and was refronted as the Lloyds printers Convalescent home from 1911-69, then became flats.  The Square was decluttered in 2001 and  is Deal's only true square.

**ALLEYS**  - See page  37

**ARCHERY SQUARE** (C, D/H. W) Formerly  called  The Archery Ground and previously Clarence road ran on its north side and Alexandra Road on the south. (D/H, W)  Tennis and croquet was played on the green, which was a subscription garden known as the Green Park.. The Dutch gabled houses were built in 1927 by Frederick Speaight, who went bankrupt. Archery Square first appears   in a 1929 directory and continuously from 1938.

**ARK LANE** (A, PH)  Previously  Noahs Ark Lane, named  after the pub which closed in 1920. Opening off this are 24 flats built in 2010 called **Sandown Place**

**BALFOUR ROAD** (B,) Arthur J Balfour, Prime Minister 1902-05

**BEACH, The** ( A, B, C, D, W)   Began in 1820s with Victorian and Edwardian detached and terraced houses. Also known as Beach Road in 1880.  A blue plaque commemorates Joseph Lister, the anti-sceptic pioneer at no. 32

**BEACH STREET**  (A, B, C, D)  Previously  Upper streate in $17^{th}$ Century.  Widened at north and south ends in 1836, and electrically lit in 1928. First mentioned in a rental survey between 1682-88. The buildings on seaward side were gradually demolished from the $19^{th}$ century until 1950 when only the Royal hotel remained. The last building to be built on that side was the Napier tavern which stood from 1855-1913. A blue plaque to Rev Stranley Treanor, the Chaplain to the local Mission to Seamen based at the Boatman's Rooms, is at the corner of Exchange Street. At no. 109 another plaque for Nathaniel Gubbins, the widely read Fleet street humourist.

**BEACONSFIELD ROAD** (A) Previously   Castle Row, after Benjamin Disraeli (1804-81) prime minister who became Lord Beaconsfield in 1876, and leading off, **Northcote Road** after Stafford Henry Northcote , Earl of Iddlesleigh , Conservative leader from 1876-85.

**BELMONT**  (A,W)  Previously Belmont Place.  Belmont Villa is at no 1. A narrow track leads to this early $19^{th}$ century terrace.

**BIRDWOOD AVENUE**  after Field Marshall Sir William Birdwood, CDC from 1935-51 who commanded ANZAC at Gallipoli. Leading off is **Delane Road**, after William Pinckard Delane Stebbing, mayor from 1947-49, who named roads himself! **Diana Gardens**   recalls his second wife;  **Grantham Avenue** refers to  WW Grantham, a Deal recorder. **Mounts Close** recalls a family of local landowners, Mounts farm being off Mill road. Upper Deal had 16 farms in 1843, only 5 by 1910. **Travers Road** is named after recorder Travers Christmas Humphries (1901-83)

**BLENHEIM ROAD** (A, D/H, partly W) Previously South Sandy Lane until 1869. The north end up to Queen St was called Wellington Place until at least 1948. Blenheim was a battle site in Bavaria where the Duke of Marlborough was victor in 1704, and he made his home at Blenheim Palace, Oxfordshire. Blenheim House was built in this street too. Leading off are **Granville Street** (A), recalling George Leveson-Gower, Earl Granville and LW from 1866, also foreign secretary. The road was accepted as a public one in 1869. **Ravenscourt Road**- built c 1910, and is being extended in 2010. The new homes are ideal for rail enthusiasts being close to the tracks. Origin of name uncertain.

**BREWER STREET** (A, C) No brewery remembered here but John Brewer, Deal's first Recorder. A soup kitchen operated here from 1851-1914 behind the Feed My Lambs school. The former Paragon Music Hall stands on the corner of Middle St.

**BRIDGE ROAD** (A, C, D/H) Previously Bridge Row Recalls a small bridge over a dyke draining the north end of the "Sea Valley"

**BROAD STREET** (A,B,C,D) Called in the press the narrowest street in 1874! We've proposed Max Broad, a baker as the inspiration here. The street was macadamised in 1844 and widened in 1926.

**BULWARK RD** (D/H) Previously Bulwark Row.(1841) Recalls bulwarks between Deal and Sandown castles. Two earth bulwarks existed -the Great and The Little Bulwark -before the castles.

**CAMBRIDGE ROAD** (A,PH, W) Previously Cambridge Street. Street is named after the Cambridge Arms pub which has been open since the early 19th century.

**CAMPBELL ROAD** (A,C (part) PH,W) Previously Campbell Street. The former Sir Colin Campbell pub was open at no.36 until 1962, and was the title of the Field Marshall made Baron Clyde in 1858.

**CANADA ROAD** (A, H/M, W) Previously Military Rd. Recalls Royal Marines Canadian campaigns as in 1812, so does the Wolfe Court flats in this road. **Canada Mews** is opposite Trafalgar Drive.

**CANNON STREET** (A) Part was originally also Cannon Road where it crossed Golf Rd. The Cannons were local landowners and a Cannons sweetshop stood nearby in College Rd. The gasworks began here in 1832 is now being redeveloped. A new road (2011) has been named **Out Downs Road**. Two more modern extensions are **North Lea** (1985) and **West Lea** (1988)

**CAPSTAN ROW** (A, D/H) Up to 32 capstans were located on beach from North St to the Time ball tower. The capstan grounds were purchased by the Corporation in the late 19$^{th}$ century.

**CAVELL SQUARE** (1934) Alderman Ernest Cavell was mayor from 1932-34, a JP and freeman . Cavell is an old Deal family; **Douglas Road** nearby commemorates Douglas Daniels OBE, Town Clerk and freeman 1952

**CENTURY WALK** (A) Previously St Andrew's Walk after 1887, as residents found postal deliveries were not getting to them at Gravel Walk . The current name marks the Church centenary in 1950. The late comedian Norman Wisdom once lived here as a child, and had a job in Liptons.

**CHAPEL STREET** (A,C,D/H) Previously Chapel Lane. A wagonway leading to St Georges, a Chapel of Ease to St Leonards, completed in 1716

**CHERITON PLACE AND ROAD** (W) Previously Cambridge Place and Vale Rd respectively. Recalls a Cheriton Cyclists' rest and temperance restaurant on the site of the Nat West bank, now a pharmacy. The names were changed by 1908. .

**CHURCH LANE** (A,) was the ancient link from Sholden Church to Southwall . Off this lane half the roads nearest Sholden church are in the parish of Sholden. In Deal **Homefield Avenue** was built in 1950; **Hunters Walk** in 1997; **Hytton Drive**; **Roman Close** (H) a rare recall of Julius Caesar's two invasions; **William Pitt Avenue**. Pitt was twice Prime Minister and LW (1792-1806)

**CHURCH PATH** (A, D) From St Leonards church to Western Road Recorded as Stoney Path in 1682 then appears on a 1769 map of East Kent by Andrews and Drury as Gravel Walk. Was Church Walk on 1873 OS map but Church Path by 1898. A bier path for a coffin and two men! Court Lodge at no.132 has a brown diamond-shaped plaque provided by Historic Houses of Kent. See also **Century Walk.**

**CHURCH STREET** (A, C, D, W) Leads past Blessed Mary church (1120) Extra drainage was provided to combat flooding. Off this street is **Green Lane (**C**)**, which probably recalls green fields in the vicinity. See also <u>Military</u> for **York and Albany Close.**

**CHURCHILL AVENUE** Recalls Sir Winston Churchill, prime minister and LW 1941-65 and made a Deal freeman in 1951. Built by Town Council in 1950s. **Dorset Gardens** is another county reference.

**CLANWILLIAM ROAD** (A) Recalls Richard Charles Francis Meade, Earl of Clanwillian and CDC from 1847-79. Developed on naval yard site from 1865 as were Ranelagh, Sondes and Stanley Roads (qv)

**COLDBLOW** (D) A descriptive name associated with Cold Blow farm

**COLLEGE ROAD** (A,B, D/H) Previously part of Lower Street but recalls Deal College founded in c. 1800. which moved away in 1911. **The Firs** was built by Steven Savage in 1984, but **King Edward Rd,** dates from Edward VII; The private **Sydcot Drive** was built in 1996 and includes Sydcot House.

**COPPIN STREET** (A,C,P) Also spelt Coppen. Joshua Coppin was Deal's first mayor in 1699, and lived in the Manor House at Upper Deal. An episode of TV's "Two Ronnies" was filmed here.

**CORNWALL ROAD** (A,B,W) Previously Park Avenue until 1890. One of a cluster of County themed road names, as is **Devon Avenue**. The builder J Turner lived at Cornwall House, 1, Cornwall Road

**COURT ROAD** (B, H, W) Formerly part of Mayers Lane. Passed near to Walmer Court, and was an ancient route to Dover via Guston. Road developed after the railway came in 1881. Leading off are **Menzies Avenue,** commemorating Sir Robert Menzies, Australian Prime Minister and LW from 1966-78, and **Reading Close** after Rufus Isaacs Marquess of Reading, LW from 1934-36 and CDC (1927-35) freeman (1934) and holder of nearly as many state posts as the 36 bungalows!

**DAVIS AVENUE** Captain Cecil P Davis was mayor from 1929-31 and 1937-38 whilst for **Frederick Road** Frederick Pearson was a director of Pearson Dorman and Long, the Mining Company.

**DEAL CASTLE ROAD** 1865 (A, B, D) Previously Castle Rd . The Castle was built in 1539-40.

**DOLPHIN STREET** (A, C, D/ H) Without an authenticated Dolphin inn nearby, a source claiming derivation from a mooring pole with iron rings to moor luggers just offshore is more plausible.

**DOVER ROAD** (A,B,D,W)  A Turnpike Rd in 1790s, but the section through Upper Walmer has been called High Street. Roads off this main highway- **Downlands** (D); **Gothic Close** (D/H) Gothic House was formerly Walmer Cottage, once owned by Lady Wollaston. It runs to a modern Mormon church. ; **Havelock Road**  possibly recalls Sir Henry Havelock who fought at the Battle of Lucknow; **Kelvedon** Road, previously Park Road East, refers to the Essex birthplace of the preacher Charles Spurgeon who was honoured in the Baptist church here (built 1904); **Kingsland Gardens**; **Lawn Road** (D/H) a gravelled cul-de-sac.  Lawn House was the home of Admiral Frederick Byng Montressor (1811-77); **The Maltings**  recalls Thompsons Brewery which closed in 1974 and was demolished in 1987-88.; **Newlands Drive** ; **Palmerston Avenue** (B) refers to Henry John Temple,  Viscount Palmerston,  Prime Minister and LW from 1861-65 ;**St Margarets Drive**  was built where William Denne the builder's home and yard was vacated in 1999; **Somerset Road,** once Park Road West, is another county name;  **Thistledown** was built in 2006; **Thompson Close** (H) -see Maltings ;  **Clifford Park** is a Caravan Park

**DOWNS ROAD**   (B, D, W) Refers to the sheltered sea lane between the shore and the Goodwin sands. Began by 1898 and extended in 1927. Opening here:-**James Hall Gardens**  recalling a Dr James Hall – a boat surgeon or doctor; **Kennedy Drive** was built after President Kennedy was assassinated.  **Owen Square** was to be called The Square by LJ Hollands & Co the builders but Admiral Sir Campbell Rich Owen  was a nineteenth century admiral who had a house where Deal Fire station is now, also a pub named after him. This kept  Owen square (1934) distinct from Archery Square.

**DRUM  HILL**  (A, B, D/M, W) A locality on Dover Road, no longer a separate address. The Drum pub existed from 1541 when the castle was built, right up to 1968 and faced the road to the castle.

**DUKE STREET** (A) No duke was involved but one of the Hayman brothers (with Robert) whose orchards were built on from 1806

**ELIZABETH CARTER AVENUE** Previously a shorter Carter Avenue and Elizabeth Avenue which were combined after the postwar prefabs were demolished..(See prefabs) Elizabeth Carter was a linguist, author and bluestocking, whose house is in South Street. Also nearby,**St Gregorys Close** Is in the saints cluster – no church

**ELLENS ROAD** Runs towards Great Mongeham and semi rural – possibly refers to farmers in locality two centuries ago. Leading north the unmade Flodden road, and several tracks across corn fields.

**ENFIELD ROAD** (A) Previously Fosters Alley Enfield House was a boarding school in Beach Street and the Enfields were 19$^{th}$ Century farmers

**ETHELBERT ROAD** (B, H) Recalls the Saxon King (560-616) who converted to Christ in 597. Off this Road **Canute Road** recalls Cnut I, Viking King of Denmark from 1014 and then of England from 1016.

**EXCHANGE STREET** (A, C, PH) The Royal Exchange pub was on the corner of Beach St. This was converted into flats after closure in 1965. Exchange St was the last to remain cobbled in this area.

**FARRIER STREET** (A, C, PH) Led to the Horse and Farrier Inn which became a private house after 1911.

---

**Terraces, Rows Cottages and Villas in 1910**
Deal library has lists of above as they existed within their roads.

---

**FREEMENS WAY** Recalls that the Royal Marines were made freemen of Deal in 1945 . Built on part of Horne's Farm in early 1960s. Edgar House (flats) recall a freeman too – James Edgar, JP. Mayor from 1910-12 and made a freeman in 1911.Opening off :-**Brambles Court Yard**; **Drum Major Drive** (M); and **Winchester Mews** – built in 2005

**GILFORD ROAD** (A) Previously a footpath known as Sheep Fold then Wolseley Terrace on an 1869 map but Gilford in County Down is also a title (Baron Gilford) in the Irish peerage, held by Earl Clanwilliam who named two roads. Gilford Road expanded from 1879 when railway extension announced. **Douglas Terrace** is recessed from Blenheim Road and dates from c.1901 The Hon Aretas Akers Douglas JP became a freeman in 1898.

**GLADSTONE ROAD** (P, W) Previously South Rope walk , being a rope walk for the naval yard but in 1866 a resident of Gladstone Villas proposed calling the whole road after William Ewart Gladstone, four times prime minister. A spur cul de sac to the railway was referred to as Blenheim Avenue as late as 1926. **Orchard Mews** recalls its former use while **Woodstock Road** has a Blenheim association being the location of Blenheim House. The north side of the road was built first.

**GODWYN ROAD** (B, H) Godwyne earl of Kent and Wessex (990-1053) was the father of King Harold. **Saxon Close** (H) repeats the Saxon theme but for **Vernon Place** see Portobello Court

**GOLDEN STREET** (A, C, D/H) Probably recalls the 17$^{th}$ century Golden Lion or Lyon inn whose location is uncertain, also possibly the Golden family. The name was in use by 1802. The road was tarred in 1883 to prevent flying chips dislodged by carts striking windows.

**GOLF RD** (A,B,D) Golf Courses were established from 1891 but the road was previously North Sandy Lane. Many newer roads open off Golf road. **Athelstan Place** refers to a Saxon King (895-940) and grandson of Alfred the Great; **Becket Close** (*); **Buckthorn Road** (DDC )refers to a plant that likes to grow on golf courses and around Sandown Castle; **Courtenay Rd**; **Fairway** (D, DDC) is another golf term ;**Langton Close** (*)– built in 2005; **Lanfranc Rd** (*) built in 2003; **Links Rd** (D) see Fairway; **Miller Close** – built in 2004 and there was a Great Mill half a mile SW of Sandown Castle and burnt down in 1896; ; **Pavilion Close** built in 2001; **Souberg Close** (M) and **Westerhout Court** (M) (or Oosterhout) are Dutch towns and the Walcheren peninsular was a scene of conflict in World War 2.with Deal a springboard, hence **Walcheren Close** (M). **Vlissingen Drive** has a second association, being twinned with Deal in 1970. (*) All were Archbishops. (DDC) Built by Dover District Council.


**GRAMS ROAD** (W) Andrew Gram was a merchant from Trondheim in Norway who built St Clare mansion c 1810. Leading off:- **Badgers Rise** where the bungalows with large gabled roof spaces copy older properties. One is named "The Sett". ; (Blake Close -see Poets Walk) and **Maple Close**

**GRANVILLE ROAD** (A, B, W) The later of two Granvilles to be a LW (in 1865) was George Levenson-Gower, Earl Granville. This Earl had a drinking fountain installed at the town hall to mark his first ten years as Lord Warden. The fountain is no longer in use. Opening off: **Knoll Place** after Knoll House, now a nursing home. and **Willingdon Place** which refers to another LW from 1936-41, Freeman Freeman-Thomas who was 1$^{st}$ marquess of Willingdon.

**GRIFFIN STREET** (A, C, PH) This led up to the Hoop and Griffin Inn, later replaced by almshouses. Also a William Griffin lived in the street. The cul de sac **George Street** was formerly Sun Alley after the Rising Sun pub which was there throughout the 19$^{th}$ century, but de-licensed in 1907. However the street and an alley are shown as George street on an 1872 map. A blue plaque on "Kingfishers"( no.13) recalls Stephen Phillips, playwright and poet, who died destitute in Deal.

**GROVE ROAD** (A, H, W) Refers to the Grove Estate, not a grove of trees

**HAMILTON ROAD** (A,W) Previously Cemetery Road up to 1952. Still passes the cemetery but now refers to Lord George Hamilton, mayor between 1909-10 and CDC from 1899-1923. Leading off :- **Stockdale Gardens** (DBC, 1934) recalls a family of Walmer landowners.

**HAROLD ROAD** (H) The last Saxon King (1022-66) killed at the Battle of Hastings, Built as link road over gardens. Leading off: **Bethany Close**

**HAWKSDOWN** (D, W) A gated private road begun in 1925. A 16th century Captain William Hawkes of Walmer Castle is given credit for the "Hawks" roads in the Book of Walmer, not falconry. This one is a gated private road, begun before the Second World War.. Also **Hawksdown Road** (D,W)

**HEAVY INDUSTRY?**

Hardly, but enough clothing, shoes, pottery, ginger beer, boats Brushes, etc. etc. were made here to justify another book .

**HERSCHELL ROAD** East and West, and Square (A, W) After Baron Farrer Herschell, CDC from 1890-99. The MP for Durham became a baron in 1886. The road was "new" in 1898. Instead of an intended Edwardian square the land was used by bungalow infill in the 1960s.

**HIGH STREET** (A,B,C,D)  Previously more accurately  Lower St up to 1879,  being lower than the shingle bank to the east.  But every town needs a High street. Widened  beyond Union Street in 1960. Pedestrianisation was extended in 1994.  **Clarence Place** was built in 1971 in faux Georgian style on the site of a demolished Georgian farmhouse.  The  Duke of Clarence, later George IV, visited Deal and was patron of the regatta begun in 1826.  Comarques at no.122  is  due  to  get  a  blue  plaque  to composer John Ireland (1879-1962)

**HOPE ROAD**  (A) Hope House, a gentleman's College, stood in Prospect Place and was advertising in 1876. The drill hall was built in 1878 but the Norman Tailyour flats opened in 1979. Tailyour was a CDC from 1972-79

**IVY PLA CE**   (A,C)  Previous names were Sand Pit Sole Lane (1782) and Folly Court up to 1870.  Now a cul de sac but once went through to Middle Street.

**KING STREET**  (A,B,PH)  Previously  Whetstone Street  after Nicholas Whetstone the  postmaster in Cromwell era,  then East India Arms Lane after the pub, then known after the Three Kings hotel (later Royal)  built c. 1720  as Three kings St, then Kings St, King's St and now King St !

**KING STREET** (A, C, W)   Origin Uncertain –confused with Deal

**KINGSDOWN  ROAD**  (D, W)   and  opening  off:-  **Hawkshill Road** (D/H) and **Hawkshill Camp Road** (D/H) *See* **Hawksdown**; Hawkshill Camp Road recalls a camp for London boys

**LIVERPOOL ROAD**  (B, P, W) Part had been known as Beach Terrace until 1866.  Robert Banks Jenkinson, Lord Liverpool was LW from 1806-1827 and also prime Minister from 1812-27.
Leading off :- **Alexandra Road** after Princess later Queen Alexandra wife of Edward VII who married him in 1863;  **Clarence Road** -see  High Street.

**LONDON ROAD** (B,D) Previously Turnpike Road. Had used Middle Deal Road but was on present track on an 1872 map. Called Upper Deal road or unnamed but named London Road by 1905. The oldest terrace (1851) faces the BP petrol station, but the Victoria Memorial Hospital is later (1924). It was here Claire Rayner trained as a nurse. Many roads open off including:- **Astor Drive** recalls Lady Nancy Astor, the first woman MP (1874-1964) and Sir John, MP for Dover and Deal. **Beechwood Avenue** had been Park Lane. Beechwood House opposite, was on the site of the present fire station. 29 flats known as Wellington Court were built for McCarthy Stone in 2004. **Bowling Green Lane** (B,PH) Previous known as Gun Lane. The Bowling Green was a mid 19th century pub used by the bowlers; **Bridgeside** (D) has 42 flats built in 1985; **Claremont Road** refers to a royal domain visited by Queen Victoria; **Cowper Rd** recalls that Lord Cowper was one of a family of Kent landowners since 13th Century. This road is on part of the Court Lodge estate and was built in 1889. **Darracott Close** The most likely Darracott was Claude Robert a JP and freeman from 1938 **Fiveways Rise** (H) recalls two adjacent houses occupied by mayor Stebbing, who was bombed out of the first in 1940; **Grange Rd** (D/H) after The Grange a house now used as Brewood Special school but was Wellington House school before that; **The Grove** (D/H) could follow a grove of trees; **Hayward Close** recalls Frederick Herbert John Hayward, mayor from 1903-04. **Manor Avenue** is built on part of the manorial lands of Deal prebend, or cathedral owned land; **Milestone Road** (H) is named after a milestone 1 mile from Lower Deal and 73 from London which stands by the Georgian Milestone House which is now a care home; **Orchard Avenue** (D/H) is a continuation of Bowling Green Lane. Flats were built there in 1952; **Sutherland Rd** was built c 1920. Formerly known as List Walk, list being a Kent name for a track. The origin is uncertain as the Sutherland hotel is not near this road; **Victoria Mews** are 42 flats built in 1985; **Warden House Mews** recalls a school used as a hospital in the Great war and moved for good to Crowborough in 1940. The house demolished in 1984.

**LORD WARDEN AVENUE** (B,H,W) has flats developed in the sixties in the grounds of Walmer Place (since demolished); **Dorset Gardens, Guilford Court, Hanover Close and Shaftesbury Court** are part of this development. Lord Dorset was a LW from the age of 20, 1708-13 and again from 1727-65; There is a Hanover Housing Association of London; Shaftesbury could be the famous philanthropist. The flats **King Charles Court** (M) were sold in 1988 by the Royal Marines for £1.2m.

**MANOR ROAD** (H) recalls a Late medieval manor (Upper Deal House) built by Deal's first mayor Joshua Coppin. The Gaunt family occupied the manor from mid $18^{th}$ century until 1910- regrettably the manor was demolished in 1965 except for the Coach House; leading off are **Bruce Close** Sir Ernest Bruce Charles, JP was made a Freeman in 1931 and was a benefactor of Deal FC; **Gilham Grove**, A private road. Previously Gilham Road was meadowland with the Gilham Sole pond; **Homestead Court**

**MARINA, The** (A. D) The name emphasises the development of Deal as a resort in the mid $19^{th}$ century. Villa residences were built by James Wise, George H Denne and AA Cavell and a new sea wall erected in 1889. Leading off :- **Dibdin Road**,(H/P) a road with no addressees, once called Foresters Opening after the pub. Charles Dibdin was a pioneer RNLI fundraiser and the name inspired a Deal and then a later Walmer lifeboat. Running inland **Hengist and Horsa roads** recall two Germanic brothers, who led the Anglo-Saxon and Jutish invaders to England in the $5^{th}$ century. JB Priestley, author of 120 books, wrote "Good Companions" at no.47, hence the blue plaque

**MARINE ROAD** (A, C, D, W) Runs Close to Royal Marine barracks and the former RM baths, replaced by the Cedars surgery.

**MARKET STREET** (A, C, D) Led to the market and the stocks – a cobbled area remains outside the public toilets

**MIDDLE DEAL ROAD** (A,B,D)Previously Back Rd Marks the boundary of Middle and Upper Deal and used as alternative route to London to avoid the track beyond Sandown castle. Shown on an 1769 map. Opening off are **Abbey Close** built in 1998 which was marketed by Abbey estate agents; **Dola Avenue**  Dola is a romantic variant of the name for Deal used by HS Chapman in "The story of Dola" the nearest actual usage being Dela in 1159; **Foster Way**; **Grace Walk** (DDC)– Grace Dobson was the wife of wartime mayor Ernest John Dobson; **Matthews Close** refers to Sir Matthew Mennes of Deal Farm; **Sheron Close** is less obvious being a mix of SHEila and RONald, forenames of the developer and his wife. (DDC) – Built by Dover District Council

**MIDDLE STREET** (A,C, D) The  Street between Lower Road and the beach. Began as a link between two Tudor castles  It has the highest ratio of listed properties in Kent's first conservation area, under the 1967 Act, and narrowly escaped much redevelopment, on top of that achieved by Hitler. There is a DDC plaque to mark Carry On actor Charles Hawtrey's memorable residence at no. 117. Queen Anne House (no. 13) is Grade II* listed, much else is grade II

**MILL HILL** (B,D/H) Previously part of Mill Road until c 1934. The smock windmill at the top near to the  waterworks  was built in 1855 and demolished in 1929.  Betteshanger colliery had opened in 1924 and much new housing was needed for the arriving miners. Nearby are **Arthur Rd** (1930) which  refers to Arthur William Lambert, freeman and mayor 1920-23 and  1927-28; **Beauchamp Avenue**  (1930) recalls William Lygon, 7$^{th}$ Earl Beauchamp who was LW from 1913; **Castle Walk**; **Celtic Road** (H) recalls the Celts who reached Mill Hill in 600 BC. **Cowdray Square**  (1929) refers to Lord Cowdray a director of Pearson, Dorman Long, the mining company. The mobile library stops here. ; **Selway Court** (1930) is not flats but a cul de sac named after the company secretary of PDL. **Yew Tree Mews** was built on the former pub's car park. A Tesco store could replace the pub.

**MILL ROAD** (A, B, D/H) Previously Flax Street and Lower Mill Lane. The Wellington flour Mill stood until c.1890 and the Chitty family were millers living in the present Linwood Youth Centre; Opening off are **Allenby Avenue** which recalls Field Marshall Lord Allenby, CDC from 1925-28 who had been High Commissioner of Egypt; Leading off:- **The Drive**; **Lower Mill Lane**; **Milldale Close** has 23 flats built in 1984**; Sycamore Drive**, **Tudor Mews** named after its half timbered look; and **Wilton Close**

**MILITARY CONNECTIONS AND SITES** (F recalls a ship lost in the Falklands War)

In the 1960s the depot could have 1,100 personnel or more. The disposal of Royal Marine property began with the sale of the Infirmary barracks for £3m and their demolition in 1988. Quite recent buildings, such as the accommodation block on the North Barracks, opened in May 1956 were demolished in 2000 but listed buildings survived, most noticeably in the East and South Barracks.

Not everyone in the Deal Charter Trustees wanted the Falklands war to be commemorated . A Falklands Road did not come about, but nor did a Belgrano road for Argentinian balance!

**Ardent Avenue** (F), **Admiralty Mews** is mainly the name for the listed 365 foot long East Barracks; **Bamford Way** refers to Major Edward Bamford (1887-1928) DSO, VC ; **Cavalry Court** is the original name of the South barracks built in 1796; **Chater Court** refers to Major-General Arthur Chater CB, CVO, OBE, 1896-1979; **Coventry Gardens** (F); **Dowell Mews** refers to Lt Col George D Dowell. VC,1831-1910 ; **Drew Lane** to Colour/sergeant John Drew, DCM, 1885; **Finch Mews** to Sergeant Norman Augustus Finch VC; **Halliday Drive** to Captain Lewis Halliday, VC,1870-1966; **Harvey Avenue** (2008) to Francis J W Harvey VC 1873-1916; **Jubilee Drive** ; **King Charles Court** -see Lord

Warden Avenue; **Maxwell Place** was built in 1991, probably refers to Major John Maxwell, beheaded by the Japanese in 1945; **Sheffield Gardens** (F); **Trafalgar Drive** after the battle; and constructed in 2010. **Wilkinson Drive** refers to Bombardier Thomas Wilkinson, VC 1832-79 ; **York and Albany Close** off Church Street. Walmer refers to the Duke of Albany's and Maritime Regiment of Foot founded by the future James II in 1664. See Also **Golf Road** for military references in road names.

**NELSON STREET** (A) Viscount Horatio Nelson 1758-1805 was a war hero and repeat visitor to Deal just a few years before this street was built from around 1811. The Zion Baptist chapel built in 1814 is now a private residence after the organ works moved to Ash.

**NEW STREET** (A, C) No longer new and this term around Middle Street is only relative, ie c. 1820.

**NORTH STREET** (A, C, D) Previously Scarborough Cat Lane after the pub which became the Globe c. 1873. A northern location is a more likely meaning than after Lord George North, CDC from1781-1802, bearing in mind South and West streets.

**NORTH BARRACK ROAD** (A, H, W) The road skirted the North Barracks built by James Johnson in 1795 but developed much later., and was straightened in 1880 by part filling in **Tapps Hole**. The latter is down steps from North Barrack Road, but after some dangerous cottages were demolished in 1892 the names appears to have fallen out of use. Now numbered in North Barrack Road.

**NORTHWALL** (A, D) Refers to a low walled embankment to regulate flooding from the marshes and first mentioned, with Southwall, in a survey dated 1617. Along the road Fairclough Homes built cottage style homes around 1984 and Tony Easton followed on with terraced homes in 1986 **Northwall Mews** was built between 1984-86 ; and **Graylen Close** in 2000. **Friends Close**, (2011) recalling the Friend family of market gardeners.

Anchor Lane – a quiet cul-de-sac off West Street

Ark Lane – the factories are no more

Dibdin Road high rise does not dwarf neighbours

Duke St towards St Andrews Church

Friends Close abuts the railway – Any High speed commuters?

Refurbished deco shelter on the Marina

Towards the Town Hall – a dignified row in mixed use

Town Hall undercroft - for buying, selling or sitting

Middle Street. A corner is shaved off for carts to pass

Ranelagh Road built over the Naval yard

The Water tower and housing with/without fireplaces

Bungalow provision near St Marys, Walmer

Building period contrasts in Church St, Walmer

Sheffield Gardens - quality build on Royal Marines land in Walmer

New Housing in Station Drive, Walmer, and left wintry Wollaston Road.

Above, Crown Court, and right Water St still retains Telegraph poles, now disappearing.

**OAK STREET** (A, C, PH) Previously Bear Pump Lane as it led to a pump on site of Town Hall, built in 1803. The car park marks the site of the Royal Oak Inn.The street stored timber for a jetty which existed from 1838-57. Off Oak street a narrow street called **The Wood Yard** had a similar function but was closed to cart traffic in 1889.

**PARK AVENUE** (D) Previously The Narrows which led though Victoria Park and became a road in the 1920s. Leading off are **Park Lea**; and **Lister Close** which has 54 flats built in 1982 on the site of a brickfield. Dr Joseph Lister, 1827-1912 was a Walmer resident famed for antisceptic surgery. Think Listerine!

**PARK STREET** (A, D) Lacks a park but the traces of the convent of the side wall of the Co-op stores recall the convent established in 1871 which stood on this street.

**PETER STREET** (P) Recalls Peter Hayman – **see Duke Street**. Established by 1813

**PILOTS AVENUE** (B, H) Recalls the pilots who protected ships from the Goodwin sands; Leading off are **Trinity Place** recalling Trinity House, authority for Lighthouses; and **Yew Tree Close**

**PORTOBELLO COURT** (A, C, H) Previously Portobello Alley from 1781 Once ran to Griffin Court off Silver St and named by a Captain who served under Admiral Edward Vernon, who captured Portobello, Panama in 1739

**PRINCE OF WALES TERRACE** (B, C, P) built on site of Naval Yard closed in 1863 and completed in 1874. Probably named after the future Edward VII who was then Prince of Wales.

**PRINCES STREET** (A, P) The name appears from 1808 and was sometimes changed to Princess by census enumerators! Possibly a reference to the Prince Regent who was standing in for George III

**QUEEN STREET** (A,B,P)  Previously Five Bells Lane after a pub renamed the Swan in 1740 but the street was renamed Queen Street after being widened in 1797. The name was not inspired by Queen Victoria, who did not become Queen until 1837. Widened again in 1927. **Queens Mews** (1986) are 31 flats.

**RANELAGH ROAD** (A, P)  built in 1865 and refers to Thomas Jones, 7th Earl Ranelagh, 1812-85 who founded a volunteer army.

**RECTORY ROAD** (B, D)  Previously called Pond Lane after a surviving pond. The Georgian Rectory for St Leonards was built by the Backhouse family. Leading off are : **Addelham Rd and Close** which refer to the name of the original village in Doomesday Book "Ad Delam"; **Bowser Close** recalls the Rev Edwin David Bowser, rector of Deal from 1919-48.; **Brenchley Avenue** is inspired by George Brenchley, town councillor in the 1930s.; **Clarkes Close** recalls two councillors not one – Lance and Emily Clarke; **Glack** Road is often referred to as The Glack, after the former house owned by the Elphinstone family used by Canadian soldiers as a hospital in the Great War; **Patterson Close** recalls another rector, Rev Robert Patterson in post from 1905-19. **Tormore Park** was built in 1980 on the playing fields of a private school founded by a Mr MacDonald who owned Tormore House in the Isle of Skye. The school was demolished in 1993 ; A blue plaque to GPR James, the historical novelist, is to be seen in Tormore Mews. **Toll Gate** refers to one closed in 1874, near St Leonards church.

**REDSULL AVENUE** (1930) (B,P)  Alderman William Henry Redsull chaired the Housing Committee, and was mayor from 1914-19 and a freeman from 1920. **Lancelot Close** has a Falklands War reference (see Ardent etc.) **Leivers Road** recalls Edward Leivers who was a manager at Betteshanger colliery who was killed in an underground accident in 1955; **Mary Road** refers to Mary, the wife of mayor Cavell (see **Cavell Square**)

**ROBERT STREET** (A, B)   Another Hayman this time - Robert (See Duke Street)

**ALL THE SAINTS!** Ss Andrews, Davids and Patricks Roads were built on the Merchants Field. St Andrews Church was built on the site of a union workhouse in 1850. The other two roads have no church, just a saint. **St Clare Road** (B, D/H, W)  refers to St Clare House used as a prep school for boys and now flats known as Leelands. There is a blue plaque to Dornford Yates, the novelist who attended St Clare's School in Grams Road, on Cedar Court.; Off this **Poets Walk** is just for Robert Bridges (1844-1930) who became poet laureate and was born in Walmer. Poets Walk was built on the site of the Convent of the Order of the Visitation of Mary, who left in 1971. Only Pugin's Catholic church remains, with a blue plaque to Robert Bridges to the rear. **Roselands** recalls one of the four Walmer estates but consists of 38 bungalows whose residents were nominated by the London boroughs. **St Georges** Rd (A) was previously  St Georges Place and north of  the chapel of ease finished in 1716. **St Georges Passage** (A,D,H) Previously  part of Chapel La (see Chapel St).; **St Leonards Road** was previously  Fagg Lane as in 1869 but now recalls the Parish Church. The estate built off this road for Utility Society began in 1938. **Charles Road** and **Leas Rd** lead off. For Charles see Bruce Close off Manor Road, Leas stands for four children of Denne the builder – Lionel, Elaine, Alan and Stanley!;
**St Martins Road** (B) is one of a saintly cluster with no church and leading off **St Nicholas Close**; **St Marys Road** (D, W) named after Walmer parish church built in 1888. Nearby is **The Shrubbery** built in the 1970s after the home of the Matthew family, owners of Walmer brewery, also modern **St Mildreds Court** after a house which stood there. **St Richard's Road** was formerly  Waterworks Road, as these were built in 1840 and the actual water tower in 1909. More houses were added in 1935 and postwar. The Anglican  church to serve Mill Hill (1934) caused the postwar name change. St Richard was Deal's first rector and was canonised in 1262. A blue plaque recalls Richard Aldington, the

novelist and poet, at no. 20-22. Off this road are recently built **Walmer Drive** and **Walmer Way**. Also off St Richards Road are **Alexandra Drive**; **Astrid Road** after the wife of John Grice Tooms, mayor from 1935-36; **Clifford Gardens**, built in c.1972 and recalling its builders; **The Conifers**; **Cross Road** an ancient trackway crossing the now busier St Richards Road; **Dossett Court** is 53 flats built in 1983; **Fairview Gardens** were built in 2001 on higher ground; **Kennett Drive**; **Lydia Road** recalls the wife of mayor Ernest Cavell; **Magness Road** remembers Lester Magness who worked at Betteshanger Colliery and was chair of the N.U.M; **Marlborough Road** in contrast refers to the Duke (1650-1722) who was Master General of Ordnance and set up gunnery ranges in Upper Deal. Three more roads use saints' names off St Richards Road – **Ss Augustine, Edmund and Francis,** .

**SALISBURY ROAD** (W) Robert Gascoyne-Cecil, 3$^{rd}$ marquess of Salisbury was three times prime minister and LW from 1896-1904. Over 20 numbers are missing on the Royal Mail Postcode finder, but this keeps odd and evens roughly opposite each other. Opening off are: **Meryl Gardens**, Meryl being a wife (or daughter) of the road's builder. Also **Windsor Court**. (Hanover Housing Association)

**SANDOWN ROAD** (A, B, C (part), D) Comes from Old English San Dun or sand hill. Sandown Terrace was built by William Betts a railway contractor in 1844. The Tudor castle took some time to fade away after sea erosion, with partial demolition recorded in 1863 and 1894. Leading off: **Albion Road** which could recall a lugger lost at sea in 1870; **Britannia Road** is probably after Britannia House. **Sandown Court** is a modern cul-de-sac.

**SILVER STREET** (A, C) Likely to be reflecting the smuggling wealth being generated although Golden Street (q.v.) may have another origin

**SONDES ROAD** (A, B, C,) built in 1865 and recalls George Mills, 5th Baron Sondes, an earl and MP for East Kent.

**SOUTH STREET** (A,B, C, D) Reflects its location. Widened in 1866 and a transport terminal ever since. Once had cobbles along the middle to help drag Mr Hayward's boats to the shore.

**SOUTHWALL ROAD** Also known as Court Lodge Rd (A, D/H) (see **Northwall**) Leading off is Minters Industrial estate

**STANHOPE ROAD** Built on the site of Hill and Sons brewery in 1902. Lady Hester Stanhope (1776-1839) was a niece of William Pitt, who improved the grounds of Walmer Castle.

**STANLEY ROAD** (A) One of the roads built on site of naval yard from 1865. The explorer Stanley was coming into public notice at the time.

**STATION ROAD** (A,B,W) Previously a leafy Broad Lane until the station was built in 1881; Opening off are:- **John Tapping Close**. This refers to a mayor of Deal from 1958-59 and a caterer to the Royal Marines. The mobile library stops here. **Mayers Road,** a corruption of Mays Road or Mays Lane as it was in 1896, and the Mays family once farmed here. The road added council housing from 1925 onwards; **Nevill Gardens**, built in 1980 recalls George Nevill Turner, who owned land here (see Walmer web) **Station Drive** facing the railway station has been developed in 2010

**STRAND**, The (A, B, C)) Previously known as Walmer Road prior to 1841 census

**SYDENHAM ROAD** (A) Known as Fishermans Row as in the 1851 census but landlords found the name made lettings difficult so the name was changed in 1887

**SYDNEY ROAD** (W) recalls John Robert Townshend, First Earl of Sydney, CDC from 1879-90 and Lord Chamberlain. Road numbering is sequential oldest right side first, but Kirk Gardens was inserted and numbered 1-18. Leading off is **Hillcrest Gardens** built in 1997 and self explanatory. At the far end is Garden Mews. The mobile library calls here.

**TELEGRAPH ROAD** (B, D/H) Recalls a chain of semaphores from Deal to Beachy Head which were replaced by the electric telegraph by 1860. The road was realigned when the railway was extended to Dover in 1881. **Forelands Square** is a late thirties development with a topographical reference. A new close of family homes named **Bevan Close** was built in 2011 by Forest Homes (Kent)-Aneurin Bevan?

**THORNBRIDGE ROAD** Thornbridge Hall in Derbyshire was the home of Charles Boot, guarantor of the National Housing Trust who mortgaged land in Mill Hill from 1934. **La Tene** and **Hallstatt Road** were named by mayor Stebbing to recognised finds from a late Iron Age culture originating in France and Austria respectively. Similarly **Quern Road** refers to a stone dating from 500 BC, a piece being set in the wall of the first house there.

**UNION ROAD** (A,H) Union Row existed by 1810, then became Union Street in 1869 with a continuation over West Street as Union Road. All is now one street. It led to the Union workhouse whose cellars remain under St Andrews Church, built in 1850. **Picketts Row** no longer exists.

**VICTORIA ROAD** (A, B, C (part) Previously Prospect Place until 1879, after Prospect House, the home of Thomas Hayward the boat builder mentioned under South Street. Developed after the naval yard closed in 1863. Queen Victoria had stayed on holiday in Walmer castle in 1841. The area was named Victoria Town, and completed in 1866. No 19 has a blue plaque to Thomas Hughes, author of Tom Brown's schooldays.

**WALMER CASTLE ROAD** Previously just Castle Road (A, D, W) Leading off are **Channel Lea**, built by Channel Securities, Folkestone around 1980; Also **Greenacre Drive, Whiteacre Dv**

**WARWICK ROAD** (W) Could Possibly recall the commander of fleet and eldest son of Earl of Salisbury rather than the Earl of Warwick , forced to land at Sandown, not Walmer, in a storm in 1563, well before the road existed!

**WATER STREET** (A, H) Site of the first wooden pipes of hollowed out elm to the marshes laid c 1699 They ran to a source later named the Railway stream.

**WELLESLEY AVENUE** The family name of the Duke of Wellington who died at Walmer Castle in 1852 after being LW since 1829. **Curzon Close** recalls that Lord George Curzon was LW from 1904-05.

**WELLINGTON PARADE** ( W) as above.; **Cecil Road** recalls Cecil Percy Davis – See Davis Avenue

**WELLINGTON ROAD** Previously Cottage Row , but renamed after Wellington Lodge. An infirmary was built here in 1863.

**WELLINGTON TERRACE** This was split in two by the Railway extension 1881 and is now Blenheim Road or Mill Road. The severed link is best viewed from the spur of Blenheim Road.

**WEST STREET** (A, D) Previously Sandy Lane & North Sandy Lane but West Street by 1806. **Anchor Lane** stands on the Anchor field where anchors were once stored and a pub called the Blue Anchor stood here until after the Great War. Sunnyside Cottages stand here; **The Avenue** is a modern cul–de-sac.

**WESTERN ROAD** (A, B, D) Leading off is St Davids Road, the Welsh patron saint, but Deal has all four! 40 sheltered housing flats named Gerald Palmby Court were built here in 1983.

**WILSON AVENUE** Recalls the son of George Wilson Daughtrey, mayor in 1945 who was reluctant to have a street called after himself. Leading off: **Little Avenue** recalls his successor Sydney Little, mayor from 1945-46.

**WOLLASTON ROAD** (A, W) Sir Gerald Woods Wollaston (1874-1957) KCB, was Norry and Ulster king of Arms. Often misspelt Woolaston. The former Boatmans' reading rooms were established here in 1873. See also Gothic Close off Dover Road.

**YORK ROAD** (A, C, W) Previously Back Lane then York Street, commenced by 1862. A popular area for fishermen due to proximity to beach. Due to poverty the residents did not always welcome improved sanitation or gas lighting, and the area was once dubbed "Blue Town" like the more famous namesake around Sheerness Docks. York is a dukedom held by the monarch's brother, James II succeeded his brother Charles.

---

**DEAL STREET** A street in Timaru, New Zealand, settled by Deal Boatmen in 1858

## ALLEYS!

These can merit a book of their own as in Whitstable, but in Deal fall into several kinds – bombed, lost, blocked off or currently extant, signed or unsigned. Most are, or were in the **Middle Street car park** area, much damaged in 1940. Many cellars filled in with rubble remain hidden under the bitumen macadam. The pre redevelopment map drawn in 1959 is more accurate than the plan showing the new layout, which can use even older names which I've noted in brackets. Facing north and clockwise there are **Blackhorse** and **North Blackhorse Alley** past the pub later known as the Strand (Steeds Alley); **Woodruffs Alley** (unsigned) adjacent to Clarks shoe shop, had been Drincobier Alley after a French soldier who set up a poulterers and later a fishmongers shop. **Cockle Swamp Alley,** recalls the cockle gatherers who cooked them in the vicinity; **Custom House Lane** which led to the second Customs House; **Oddfellows Alley,** skirting the former Oddfellows Hall (1890) **Five Step alley** is stepless now (Earnshaws Passage); *Short Street* and skirting the Quarterdeck are **Crispin Alley** (unsigned) and **Tucks Alley** (curtailed) Passing the Beachbrow Hotel rebuilt in 1808 are **Primrose Hill** and **Primrose Alley** which originally began opposite the rear of the present library. This name recalls the Rev Daniel Primrose, whose wife famously records he died of lethargy. **Bakers Alley** (unsigned) ran south of the Oddfellows Hall and like **Coach Yard,** which bordered the west side of the space, is no longer a street defined by buildings .Victory Alley near Short street has gone like the Old Victory pub, bombed in 1940.

Some alleys are not much more than uninhabited brick canyons, but these can be found. **Crown Court** (C) off Middle Street south of Broad Street, **South Court** off South Street which was more aptly called Jews Harp alley because of its odd shape, the usage dating back to the early 18th century. **Kings Arms alley** now blocked off and with an unofficial sign board recalls that pub and faces Christmas House in Middle Street. **Port Arms alley** skirts the 17th century pub. **George Alley** was once known as Bents Passage. **Sharps Alley** (to Royal Hotel was Chapel Alley in 1915)

Dozens of alleys are extinct so the walker may find nothing to look at.

Jezebel Alley between Robert St and Nelson St and scene of a bitter dispute in 1878 when a resident blocked it off.
Smiths Folly, 203 High St next to Lloyds Court is no more. .
Ticklebelly Alley leading to St Patricks Road does exist in part at the side of Sainsburys, but is unlikely to get a Council nameplate,

Strangle Alley was really Albert Alley between Albert Road and London Road, passing allotments and railway work-shops but nicknamed thus after a conductress was mugged of her takings on her way to the bus depot.
- Drink of Beer Alley is an anglicized Drincobier Alley (q.v.)

## CYCLE !

At least two mileposts funded by the Royal Bank of Scotland mark the route of the National Cycle Network. One is on the promenade in front of Deal Castle, the other on Golf Road on the Sholden boundary. I've seen 19 German cyclists discovering Walmer in one day!

## BUILDINGS OF SPECIAL HISTORIC OR ARCHITECTURAL INTEREST – WHERE TO FIND THEM

The Department of the Environment issued a checklist of what listed buildings exist in Deal and Walmer streets with Beach and Middle Streets having the most to see. Also Listed grade II buildings are found in Alfred Square, Archery Square, The Beach, Blenheim Road, Brewer Street, Broad Street, Chapel Street, Church Path, Church Street, College Road, Coppin Street, Crown Court, Dolphin Street, Dover Road, Exchange Street, Farrier Street, Golden Street, Grams Road, Griffin Street, High Street, Liverpool Road, London Road, Manor Road, Market Street, Middle Deal Road, Mill Road, Nelson Street, New Street, North Street, Oak Street, Portobello Court, Princes Street, Queen Street, Rectory Road, Sandown Road, Silver Street, South Street, Stanhope Road, The Strand, Union Road, Victoria Road, Walmer Castle Road, Water Street and Wood Yard.

## FLAT ROOFS
Some good examples can be seen in Leas Road

## HOLES!

Apart from **Tapps Hole** (see North Barrack Road), **Pope's Hole** later Popes Court, has attracted interest. There was a cottage owned by John Pope, and a hole nearby was a source of lime far back enough to be used for the foundations of St Leonards Church. Smugglers found the hole useful too. Cottages here were joined up and converted into tea rooms, which lasted until 1935. A Georgian style house was built for Lord Leslie Hore-Belisha (1893-1957), whose winking orange beacons were first introduced in 1934 and still help people cross the road.

## POSTCODES

Alphanumeric postcodes were introduced in Norwich from 1959. The minister responsible was Ernest Marples, in whose vast Ministry of Transport the author was destined to work at the time of the feared Beeching report on the railways. The whole of the UK was given postcodes by 1974, but none exist in Eire.

## PREFABS!

These were built post war in Carter Avenue , College Road, Marlborough Road, Northwall, Owen Square, Pilots Avenue and Trinity Place. By 1968, all had gone or been replaced by Llewellyn Quick builds, but some in Pilots Avenue have been encased in brick and concrete skins with new roofs. Other council built property is to be found in these areas, as well as Mill Hill. (Source David Collyer)

## PROMENADES

The present custom of walking by the sea was made possible by gradual extension, beginning with the South Esplanade in 1841. The prom in front of Victoria Town opened in October 1879, and by 1894 four miles of asphalted prom ran from Sandown Castle to North Barrack Road, the Walmer part being agreed by the Walmer Board in April,1893. There was a coastguards enclosure then the promenade continued to Walmer Place. But tfishermen objected to their construction, as it restricted beach space for their luggers.

## THE MILL HILL AREA
Deal Borough Council were initially reluctant to take on the housing construction needed for the newly arriving miners, on financial and social grounds. A subsidiary of Pearson, Dorman and Long, the mining company, (Snowdown & Betteshanger Tenants Ltd) built Arthur Road, Beauchamp Avenue, Cowdray Road and square, Davis Avenue, Douglas Road, Frederick Road, Mary Road, Mill Hill, Redsull Avenue, and Selway Court. The First National Housing Trust of Sheffield built Astrid Road, Cavell Square, Celtic Road, La Tene, Lydia Road, Quern Road, part of St Richards Road and Thornbridge Road. The Deal Borough Council built Brenchley Avenue, Douglas Rd, Little Avenue, Pilots Avenue, part of St Richards Road and Wilson Avenue. A street name for any of these may recall a sympathetic councillor.

## RHUBARB, RHUBARB
The origin for the name for Timperley Close (off Church Lane) is indisputable. A field of Timperley rhubarb is close by!

## ROPES
Space was needed to spin out hawsers before chain cables for anchors. Ropes were spun in Gladstone and Sandown roads.

## STREET NUMBERING
At one time houses had no numbers, rather like in Albania today, When they began to appear, several houses could allocate themselves the same number, and so caused much confusion, until the authorities got a grip. By 1890 it was thought that the style of name plate should be similar in Deal and Walmer, and Deal Council decided to number streets progressively, with odds and evens roughly opposite each other. Modern infill may need to be given A or B suffixes..Apparently, the Royal Marines origin of Kings Charles Court means those flats off Lord Warden Avenue are numbered in a different way from neighbouring flats.

## PUBLIC CONVENIENCES

These now teeter on the edge of closure in the age of austerity, but the Telegram mentions one in 1858 on the North Esplanade near Pilot House, and three urinals for men only were approved in Dec 1864. One sited behind 94 Beach Street was said to have depreciated the value of a house nearby! One site, Alfred Square has no loos now, but Deal's present pair are sited behind the South Street bus waiting room, and at the foot of King Street. Confusion reigned a few years ago when the gents and ladies were reversed! The other facility is at the south end of Victoria Park. Upper Walmer is bereft, but the Marke Wood recreation ground has toilets, more are at the sea end of Granville Road. Lower Walmer has well used loos on the Marine Road opposite the Cedars Surgery. Interestingly older toilets further south are now in use as the Sea Café. Not everyone realises that the floral roundabout at the sea end of Broad St conceals underground toilets that would be difficult to access in today's traffic.

## STREET CLEANING

Nowadays we can expect a street cleaner picking up stray litter, perhaps once a week in a quieter street. But up to 1858 one donkey pulling a cart coped with the entire rubbish collection, whilst by law residents were required to sweep the pavements outside their homes between 6 and 9 am. Gulley cleaning is now essential if the drainage chambers are not to clog up with leaves and debris and become useless.

## STREET FURNITURE

The public can have an influence on this, as the District Council have a range of commemorative wooden benches with name plates, or plaques. Plenty in situ!

## STREET LETTERING

Traffic signs generally use a font known as Transport Heavy, which was developed in the late fifties. Street signs locally use a more delicate font known as MOT Serif, or Kindersley, after the designer.-, David Kindersley who hated Transport Heavy. This slightly serif style all in capitals is not mandatory, but generally suits historic towns. Older styles can still be seen, as in the Upper Gladstone Road sign near the funeral parlour.

## UTILITIES

**Water** came first in 1699 with the water conveyed from the North stream in a pipe made of bored out logs. The pipes ended in Water street via a mill in Northwall. The waterworks in Upper Deal opened in 1840 (see St Richards Road)  The **gasworks** began in 1832 with a small gasholder which lit the gas lamps. A bigger gasometer came in 1859. Issues about more economic street lighting aired on the KCC website are nothing new – in 1878 the Deal lamplighters were told not to light up on moonlit nights! The site off Cannon Street was extensive until 1956 when operations and staff were moved to Dover. After an expensive decontamination of the site, regeneration is now well under way. **Electricity** was proposed for Deal from 1882, although some councillors were shareholders in the gas company! As before, proposals affected street lighting first, with an offer for 187 street lamps along High Street up to Albert Square. The Marina was electrically lit in 1928. **Telephone** wires first linked Dover and Canterbury, and in 1888 the South of England Telephone Co. wanted to connect Deal up via Wingham. Wires were erected down the old Sandwich road to Deal College and a Mr Steed Bayly hosted the first telephone exchange in his store. One of the first lucky customers was a  Dr. Bruce Payne, who was the  Port and Urban Sanitary Officer. The current exchange faces Deal station, but was formerly at 25, Stanhope Road between 1929 and 1969.

Below, Heating sources two centuries apart – the solar heated North Deal Centre, and a gasholder in distance.

Wait here for the 13, 13A, 14, 82, 541, 542, 544, 593 and free bus from Tesco

46

Lydia Road is named after mayoress Lydia Cavell, wife of Ernest Cavell. Below, the Duke of Wellington.

No street is complete without some furniture.

## STREET FINDER

Key streets and those in the historic heart of Deal are indexed separately, as these are likely to be most visited. Other streets are listed under a key road they lead off .e.g London Road

**A** Addelam Close and Rd – see Rectory Rd; Admiralty Mews – Military; Albion Rd -Sandown Road; Alexandra Dv – Saints; Alexandra Rd – Liverpool Rd; Allenby Av – Mill Road; Anchor Lane – West St; Ardent Av – Military; Arthur Rd – Mill Hill; Astor Dv – London Rd; Astrid Rd – Saints; Athelstan Place – Golf Rd

**B** Badgers Rise – Grams Rd; Bamford Way – Military; Beauchamp Av – Mill Hill; Becket Close – Golf Rd; Beechwood Av – London Rd; Bethany Close – Harold Rd; Bowling Green Lane – London Rd; Bowser Close – Rectory Rd; Brambles Court Yard – Freemens Way; Brenchley Av – Rectory Rd; Bridgeside – London Rd; Britannia Rd – Sandown Rd; Bruce Close – Manor Rd; Buckthorn Close – Golf Rd

**C** Canute Rd - Ethelbert Rd; Castle Walk – Mill Hill; Cavalry Court – Military; Celtic Rd – Mill Hill; Channel Lea – Walmer Castle Rd; Charles Rd – Saints; Chater Court – Military; Clarence Place – High St; Clarence Road – Liverpool Rd; Claremont Rd – London Rd; Clarkes Close – Rectory Rd Clifford Gdns – Saints; Clifford Park – Dover Rd; The Conifers – Saints; Courtenay Rd – Golf Rd; Cowdray Square and Road – Mill Hill; Cowper Rd – London Rd; Cross Rd – Saints; Curzon Close – Wellesley Av

**D** Darracott Close – London Rd; Delane Rd – Birdwood Avenue; Devon Avenue – Cornwall Road; Diana Gardens – Birdwood Av; Dibdin Rd – The Marina; Dola Av – Middle Deal Rd; Dorset Court – Lord Warden Av Dorset Gdns – Churchill Av; Dossett Court – Saints; Douglas Rd – Cavell Sq; Douglas Tce – Gilford Rd; Dowell Mews, and Drew Lane – Military; Downlands – Dover Rd; The Drive – Mill Rd; Drum Major Dv – Freemens Way

**F** The Fairway – Golf Rd; Fairview Gdns – Saints; Finch Mews – Military; The Firs – College Rd; Fiveways Rise – London Rd; Flodden Rd – Ellens Rd; Forelands Sq – Telegraph Rd; Foster Way – Middle Deal Rd; Frederick Rd – Davis Av;

**G** The Garth – Albert Rd; Gaunt Close – Manor Rd; George St – Griffin St; Gilham Grove – Manor Rd; Glack Rd – Rectory Rd; Gothic Close – Dover Rd; Grace Walk – Middle Deal Rd Grange Road – London Rd; Grantham Av – Birdwood Av; Granville St – Blenheim Rd; Graylen Close – Northwall; Greenace Dv – Walmer Castle Rd; Green la – Church St The Grove – London Rd; Guilford Court – Lord Warden Av

**H** Halliday Dv – Military; Hallstatt Rd – Thornbridge Rd; Hancocks Close – Lord Warden; Harvey Av – Military; Havelock Rd – Dover Rd; Hanover Close – Lord Warden Av Hayward Close – London Rd; Hengist Rd – The Marina; Hillcrest Gdns – Sydney Rd; Homefield Av – Church Lane; Homestead Court – Manor Rd; Horsa Rd – The Marina; Hunters Walk – Church Lane; Hytton Drive – Church Lane.

**J**  James Hall Gdns – Downs Rd;  John Tapping Close – Station Rd;  Jubile Dv – Military

**K**  Kelvedon Rd – Dover Rd; Kennedy Dv – Downs Rd;  Kennett Dv – Saints;  Kings Edward Rd – College Rd;  Kingsland Gdns – Dover Rd;  Kirk Gdns - Sydney Rd;  Knoll Place – Granville Rd

**L**  La Tene – Thornbridge Rd;  Lancelot Close – Redsull Av; Lanfranc Rd – Golf Rd;  Langton Close – Golf Rd;  The Lawn – Dover Rd;  Leas Rd – Saints;  Leivers Rd – Redsull; Av;  Links Rd – Golf Rd; Lister Close – Park Av;  Little Av – Wilson Av; Lower Mill Lane – Mill Rd; Lydia Rd – Saints

**M**  Magness Rd – Saints; The Maltings – Dover Rd;  Manor Av and Close – London Rd;  Maple Close – Grams Rd;  Marlborough Rd – Saints;  Mary Rd – Redsull Av;  Matthews Close – Middle Deal Rd;  Maxwell Place – Military;  Mayers Rd – Station Rd; Menzies Av – Court Rd;  Meryl Gdns – Salisbury Rd;  Milestone Rd – London Rd;  Milldale Close – Mill Rd;  Miller Close – Golf Rd;  Mounts Close – Birdwood Av;

**N**  Nevill Gdns – Station Rd   Newlands Dv – Dover Rd; North Lea – Cannon St; Northcote Rd – Beaconsfield Rd

**O-P**  Orchard Av – London Rd; and Mews – Gladstone Rd;  Out Downs Rd – Cannon St;  Owen Sq – Downs Rd   Palmerston Av – Dover Rd;  Patterson Close – Rectory Rd;  Pavilion Close – Golf Rd;  Poets Walk - Saints

**Q -R**  Queens Mews – Queen St; Quern Rd – Thornbridge Rd; Ravenscourt Rd – Blenheim Rd;   Reading Close – Court Rd; Roman Close – Church Lane; Roselands – Saint.

**S** See the Saints section, with most at the Sholden end of St Richards Road, or near St Andrews Church; examples are: St Gregorys Close – Elizabeth Carter Av; - St Margarets Dv – Dover Rd; Sandown Place – Ark Lane; Saxon Close – Godwyn Rd; Selway Court – Mill Hill; Shaftesbury Court – Lord Warden Av; Sheffield Gdns – Military; Sheron Close – Middle Deal Rd; The Shrubbery – Saints; Somerset Rd – Dover Rd; Souberg Rd – Golf Rd; Station Dv – Station Rd; Stockdale Gdns – Hamilton Rd; Sutherland Rd – London Rd; Sycamore Dv – Mill Rd; Sydcot Dv – College Rd.

**T** Thistledown – Dover Rd; Thompson Close – Dover Rd; Timperley Close - Church Lane; Tollgate – Rectory Rd; Tormore Mews/ Park – Rectory Rd; Trafalgar Rd – Military; Travers Rd – Birdwood Av; Trinity Place – Pilots Av; Tudor Mews – Mill Rd.

**V-W** Vernon Place- Godwyn Rd; Victoria Mews – London Rd; Vlissingen Dv – Golf Rd; Walcheren Close – Golf Rd; Walmer Gdns and Way – Saints; Warden House Mews – London Rd; West Lea – Cannon St; Westerhout Close – Golf Rd; Whiteacre Dv –Walmer Castle Rd; Wilkinson Dv – Military; William Pitt Av – Church Lane; Willingdon Place – Granville Rd; Wilton Close – Mill Rd; Winchester Mews – Freemens Way; Wood Yard – Oak St; Woodstock Rd – Gladstone Rd.

**Y** Yew Tree Close – Pilots Av; Yew Tree Mews – Mill Hill; York and Albany Close – Military; Young Close – Church Lane.

# ACKNOWLEDGMENTS

This book has been delayed by locked doors when Deal Library was extensively refurbished. The East Kent Archives face relocation to Maidstone, and the Maritime Museum remains closed. However the first two sources were essential, with Sandwich town archives and Dover library also helpful in providing street directories. A mish mash of records to trawl through (as one archivist put it) included hand written minutes of varying clarity, and old maps provided evidence of street growth, which continues today. Past historians have turned briefly to street names, Gertrude Nunn, LW Cozzens, Barbara Collins and several articles in the East Kent Mercury being useful sources. Mr Ebbutt. (Military roads). The Laker, Pritchard and Pain histories of Deal also must be consulted. Wartime destruction is recorded by David Collyer in Deal and District at War, 1939-45. (1995) Alan Percy Walker has painted Deal's alleys (Joem Gallery)

Deal Town Council and Wikipedia have supplied biographical data and details of Deal mayors. Lords Warden at Walmer Castle and Captains of Deal Castle came both from the internet and site visits. Brian Groser of the Deal Society helped me over the siting of blue plaques. Several individuals shared their local knowledge generously, including pub historians Michael Rogers and Steve Glover, Alan Buckman, Chris Marshall , Pat Streater and Sue Travis. The photographs are almost entirely mine, except the portrait of Lydia Cavell, courtesy of Dover Museum. The Duke of Wellington portrait is owned by Walmer Parish Churches. My own colour photos are not intended to replicate the standard central views in other books, but to reveal the architecturally varied streetscape found in Deal and Walmer. This was first made apparent to me in the excellent Walmer Design Statement, published in 2006. Two years later, the Campbell Lungair Deal Map provided a colourful portrait of the town and highlighted the key buildings, new and old, also when certain areas were developed.